EPIC

Action and adventure collide in **EPIC**. Plunge into a universe of powerful beasts, hair-raising tales, and high-speed excitement. Astonishing explorations await. Can you handle it?

This edition first published in 2021 by Bellwether Media, Inc.

No part of this publication may be reproduced in whole or in part without written permission of the publisher. For information regarding permission, write to Bellwether Media, Inc., Attention: Permissions Department, 6012 Blue Circle Drive, Minnetonka, MN 55343.

Library of Congress Cataloging-in-Publication Data

LC record for Virtual Reality available at https://lccn.loc.gov/2019059307

Text copyright © 2021 by Bellwether Media, Inc. EPIC and associated logos are trademarks and/or registered trademarks of Bellwether Media, Inc.

Editor: Kieran Downs   Designer: Josh Brink

Printed in the United States of America, North Mankato, MN.

TABLE OF CONTENTS

TIME TO RACE!	4
WHAT IS VIRTUAL REALITY?	6
HOW IT WORKS	8
HISTORY	12
TECHNOLOGY OF TOMORROW	18
GLOSSARY	22
TO LEARN MORE	23
INDEX	24

Time to Race!

The other **kart** is catching up. You twist through sharp turns. Finally, you reach the finish line. Winner!

You take off your virtual reality **headset**. It made you feel like you were really racing!

VR Karts

What is Virtual Reality?

Virtual reality is a fake world made by a computer. It can be controlled by the person using it. It is often called VR.

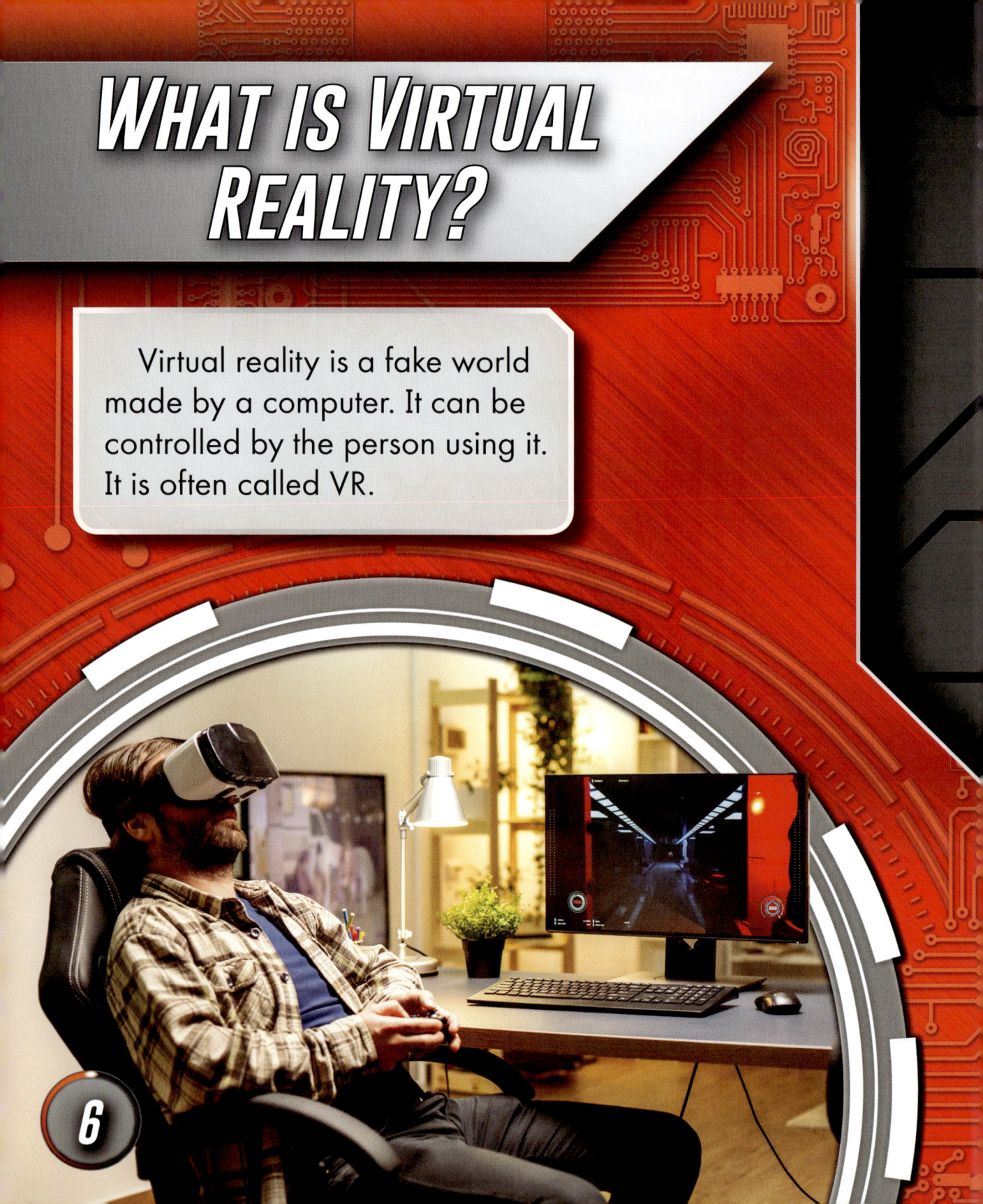

Who Uses It?

video game designers

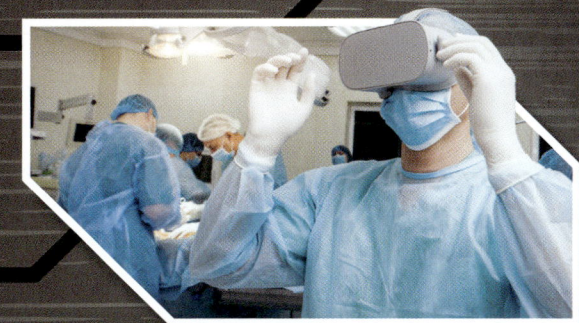
doctors

astronauts

soldiers

VR is found in many devices. Some smartphones and video games have VR!

How it Works

VR uses special **hardware**. The headset is the most important piece. A screen inside shows the virtual world. Headphones and controllers let users hear and move in the world.

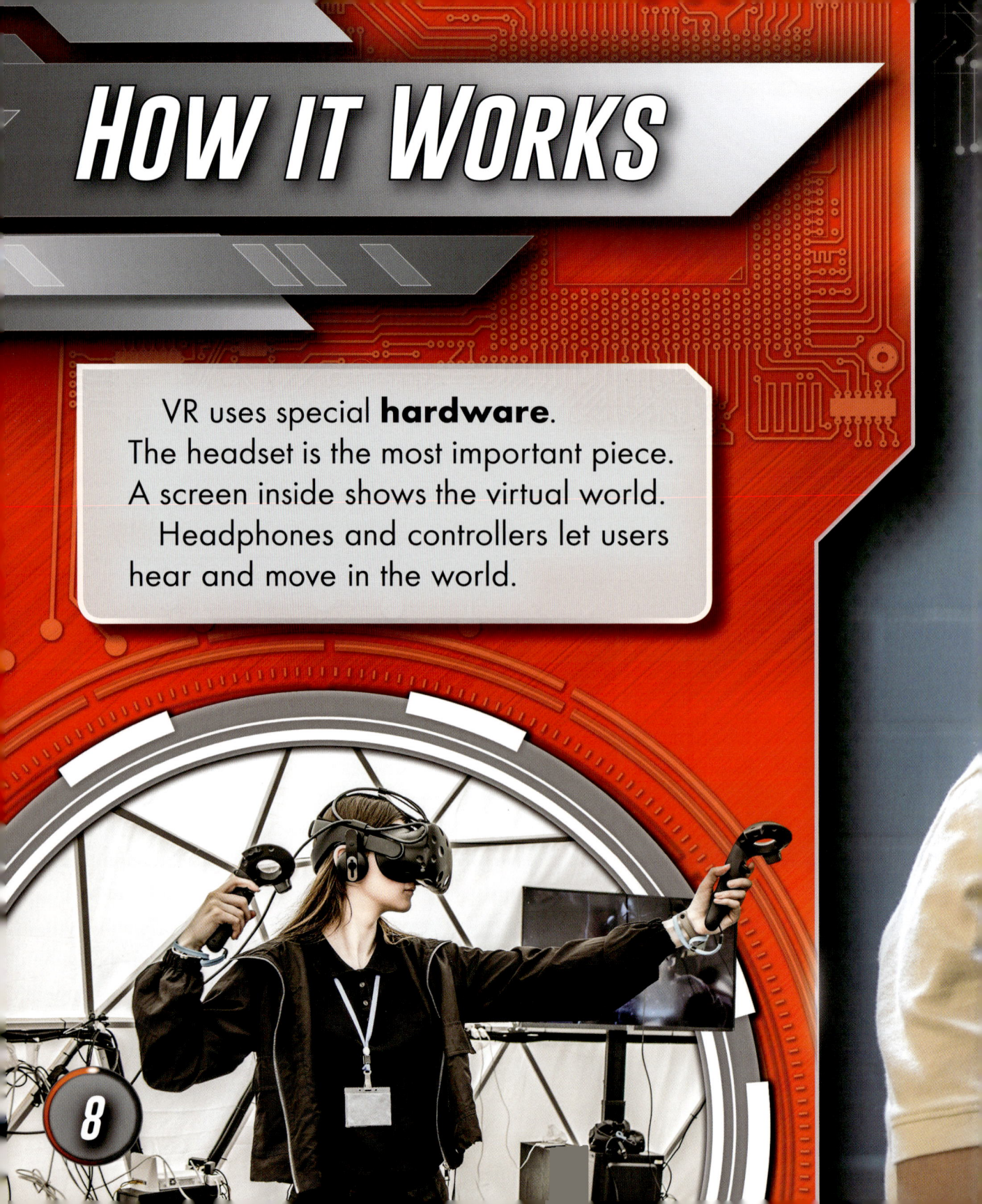

VR hardware has cameras and **sensors**. These track the user's movements. They also tell the computer where the user is looking. The computer has **software** that **generates** the world. It changes when the user looks around!

PlayStation VR

How PlayStation VR Works

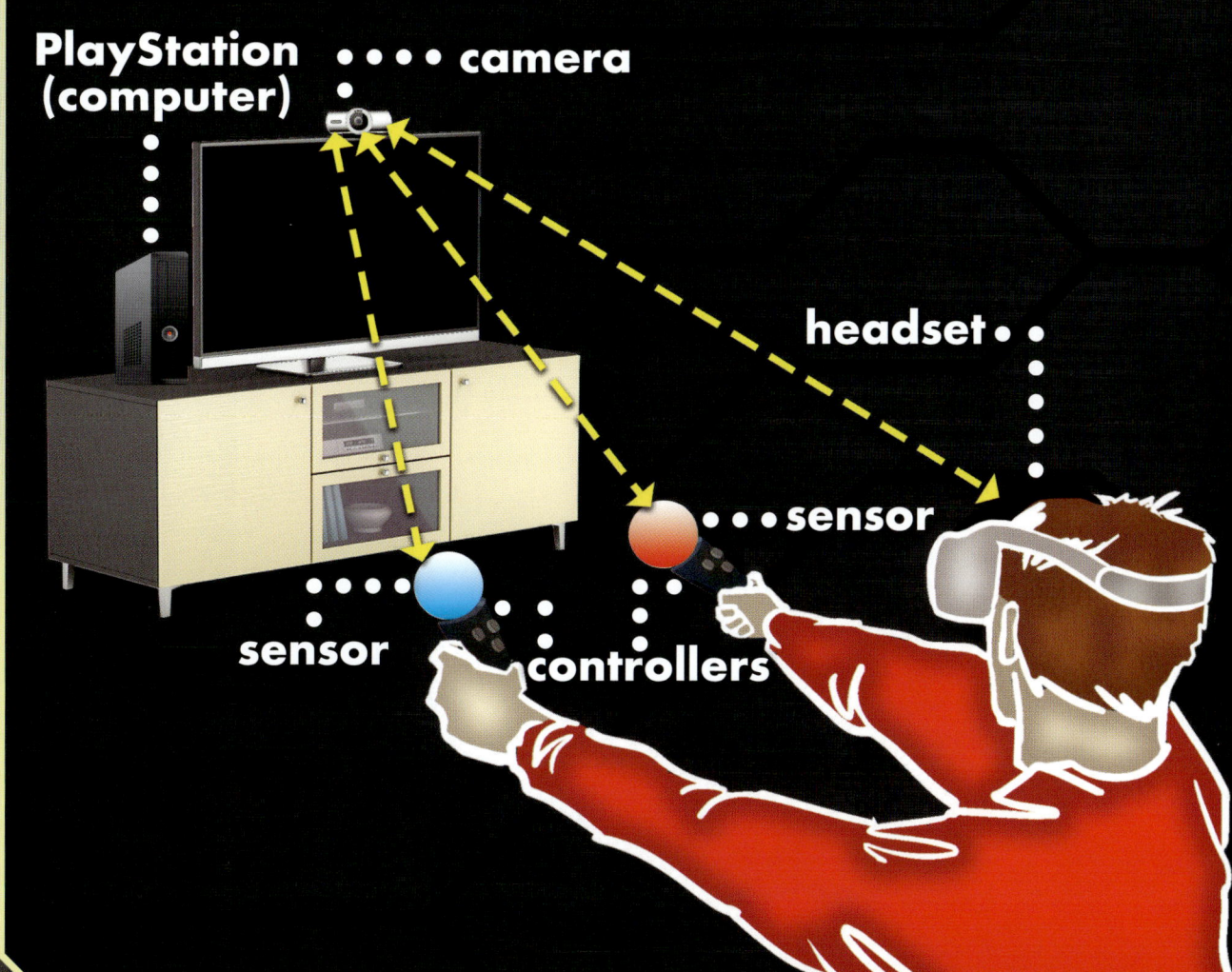

HISTORY

The first VR machine came out in 1962. It was called the Sensorama. Later, the Sword of Damocles used a headset. It put fake pictures on top of the real world. This was the first use of **augmented reality**.

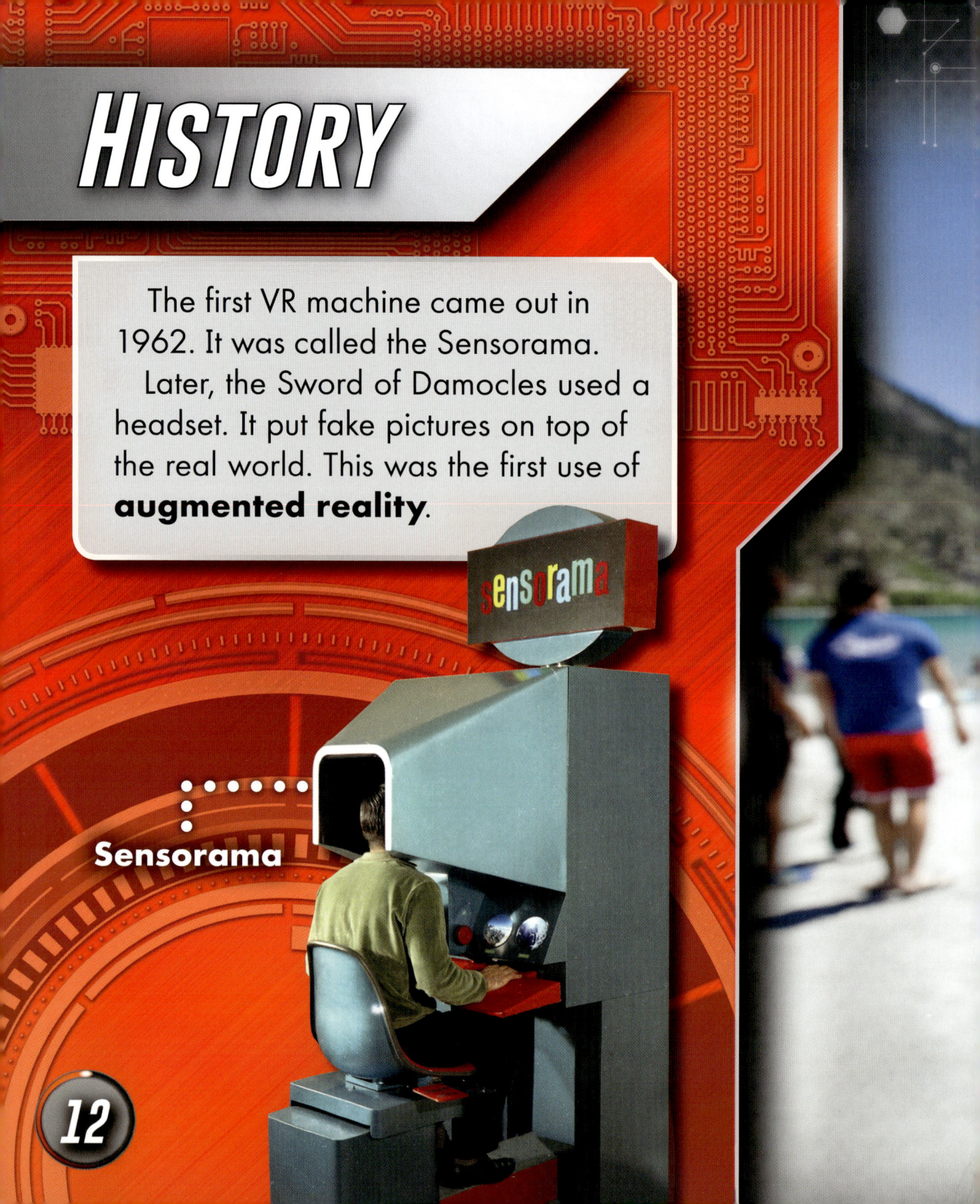

Sensorama

POKÉMON GO

Pokémon Go uses augmented reality. It shows Pokémon in the real world.

Virtual Reality Timeline

1962
Morton Heilig introduces the Sensorama, an early VR system

1985
The VIEW headset uses CGI to train astronauts

1968
Ivan Sutherland creates the Sword of Damocles, the first VR system that used a headset

By the 1970s, VR was used to train pilots and astronauts. Flight **simulators** helped them learn their jobs!

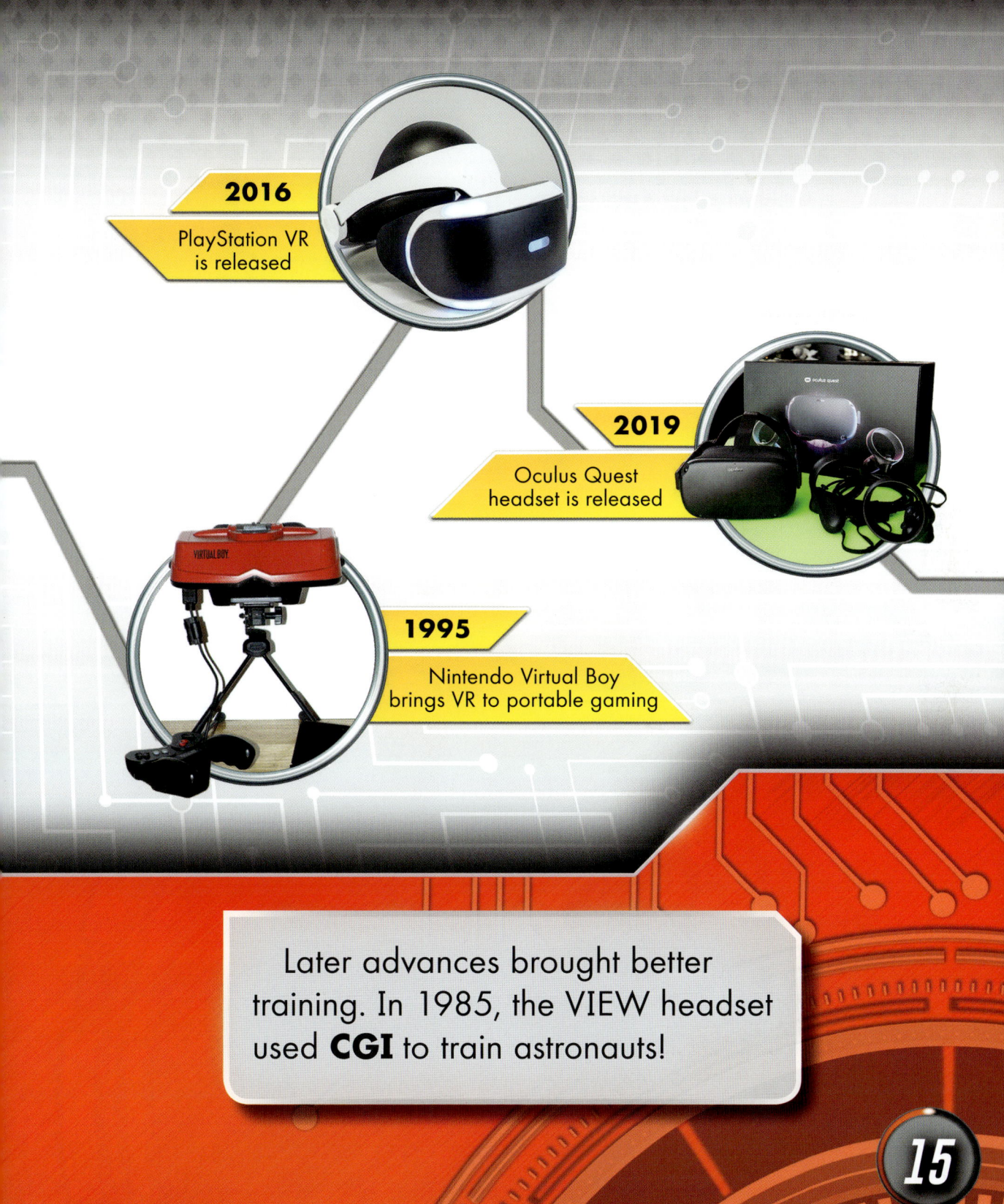

2016 — PlayStation VR is released

2019 — Oculus Quest headset is released

1995 — Nintendo Virtual Boy brings VR to portable gaming

Later advances brought better training. In 1985, the VIEW headset used **CGI** to train astronauts!

Today, VR has even more uses. Doctors use simulators to train for surgeries. Museums use VR to help people see faraway places!

IRON MAN

PlayStation released *Iron Man VR* in 2020. The game lets people pretend they are flying!

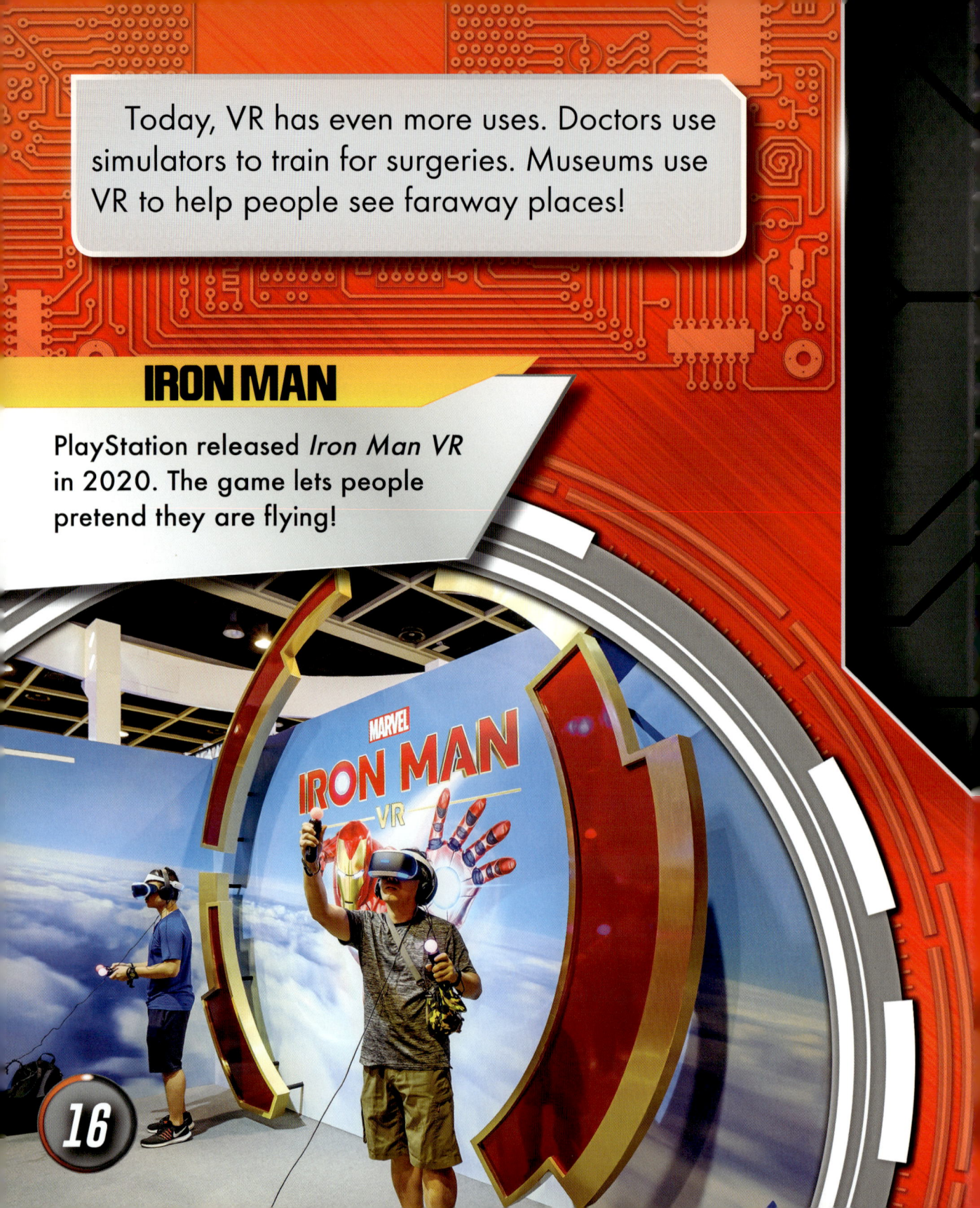

How Many Users?

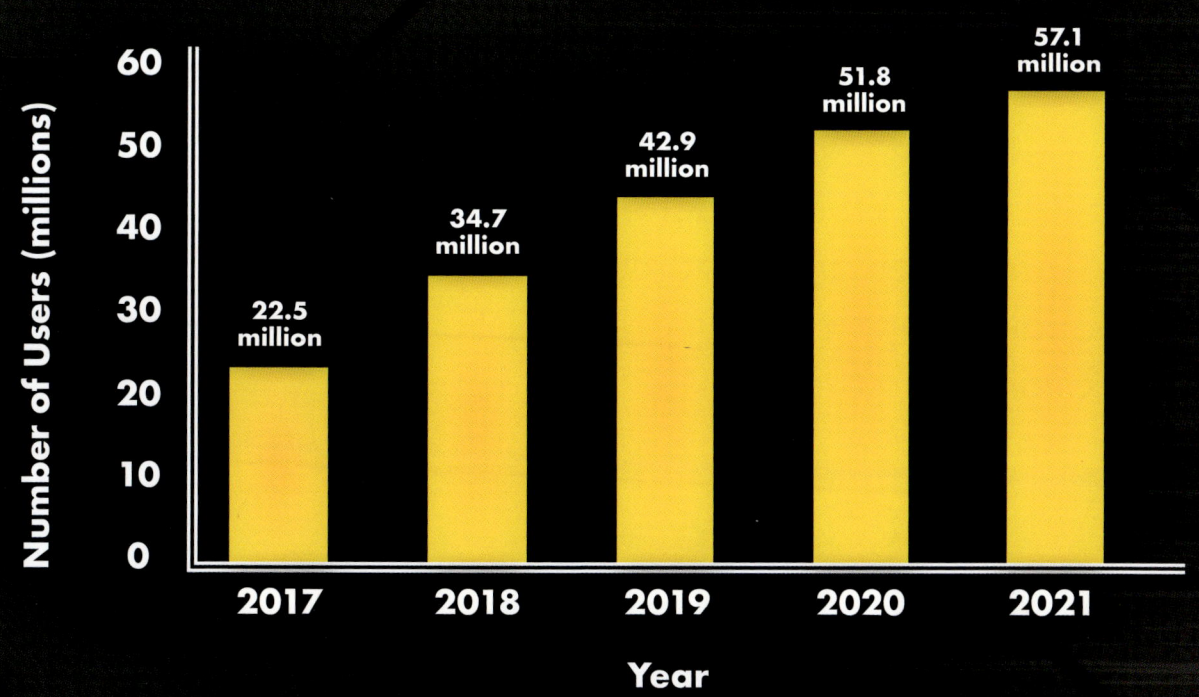

The number of VR users keeps growing.

There are many VR games, too. In 2016, the PlayStation VR and Oculus Rift brought VR into homes!

Technology of Tomorrow

VR has an exciting future. Advances could make VR even better.

Companies are researching ways to make **haptic** suits more realistic. VR will feel more real!

haptic glove

TRAINING SUIT

In 2018, Tesla showed off the haptic TESLASUIT. The company made it to help train people for dangerous jobs such as space travel and emergency response.

TESLASUIT haptic suit

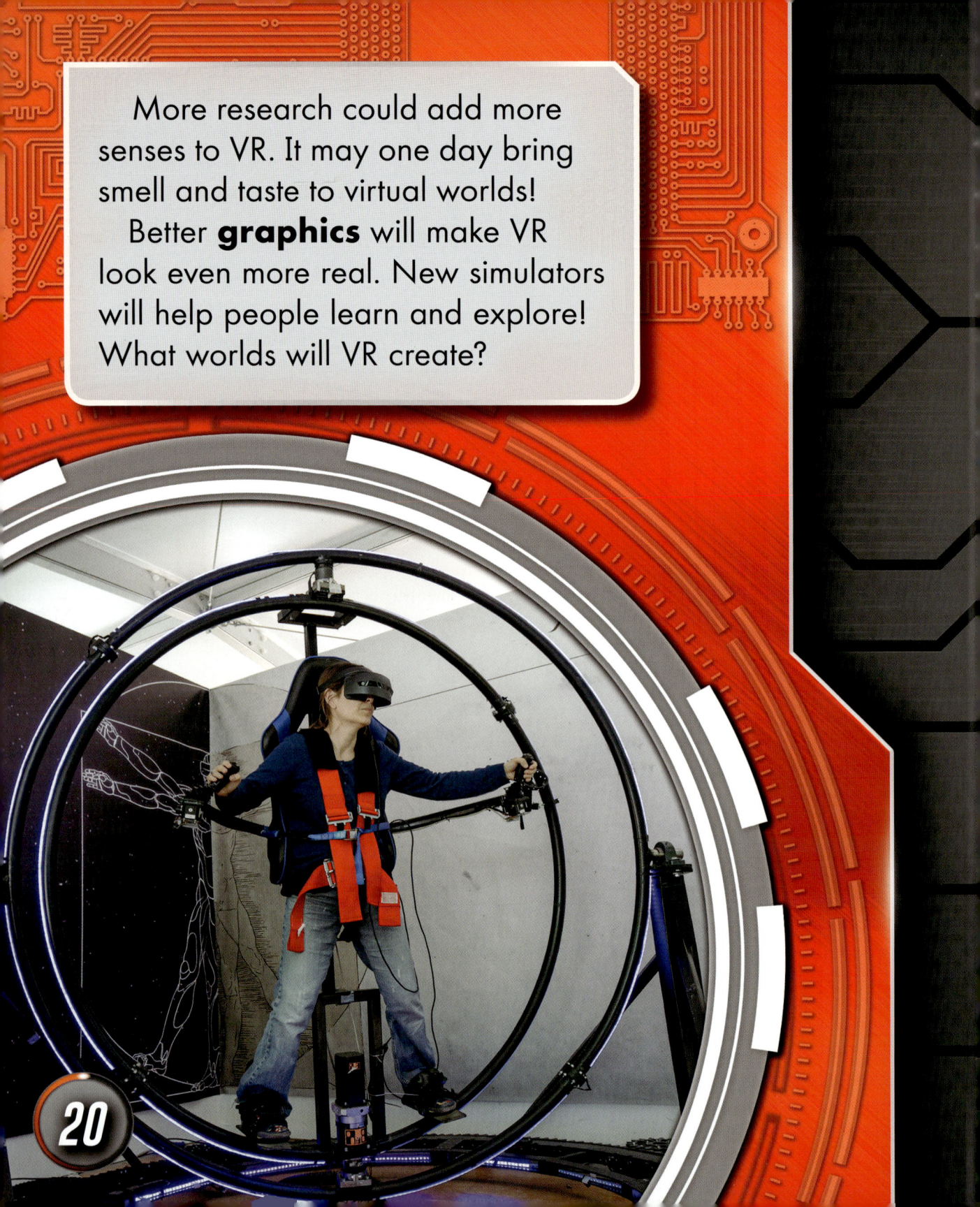

More research could add more senses to VR. It may one day bring smell and taste to virtual worlds! Better **graphics** will make VR look even more real. New simulators will help people learn and explore! What worlds will VR create?

Pros and Cons

Pros

safer training

fun

health uses

Cons

costs a lot of money

can make people feel sick

technical issues

GLOSSARY

augmented reality—technology that places pictures or writing over images of the real world

CGI—artwork created by computers; CGI stands for computer-generated imagery.

generates—creates

graphics—computer-made images that make up a VR world

haptic—related to touch; haptic clothing makes wearers feel like they are being touched.

hardware—physical pieces of computers or other technology

headset—a helmet, visor, or other item that people wear on their heads to experience VR

kart—a small, lightweight vehicle used for racing

sensors—small devices that measure something physical and react to it

simulators—programs that use virtual reality to teach people how to do things

software—computer programs

To Learn More

AT THE LIBRARY

Eboch, M. M. *The Future of Entertainment: From Movies to Virtual Reality*. North Mankato, Minn.: Capstone Press, 2020.

Rathburn, Betsy. *Drones*. Minneapolis, Minn.: Bellwether Media, 2021.

Small, Cathleen. *Using VR in Gaming*. New York, N.Y.: Cavendish Square, 2020.

ON THE WEB

FACTSURFER

Factsurfer.com gives you a safe, fun way to find more information.

1. Go to www.factsurfer.com.

2. Enter "virtual reality" into the search box and click 🔍.

3. Select your book cover to see a list of related content.

Index

advances, 15, 18, 20
astronauts, 14, 15, 19
augmented reality, 12, 13
cameras, 10
CGI, 15
computer, 6, 10
controllers, 8, 9
doctors, 16
graphics, 20
haptic suits, 18, 19
hardware, 8, 10
headphones, 8, 9
headset, 4, 5, 8, 12, 15
history, 12, 13, 14, 15, 16, 17
Iron Man VR, 16
museums, 16
Oculus Rift, 17
pilots, 14

PlayStation VR, 10, 11, 16, 17
Pokémon Go, 13
pros and cons, 21
screen, 8
senses, 20
Sensorama, 12
sensors, 10
simulators, 14, 16, 20
smartphones, 7
software, 10
Sword of Damocles, 12
Tesla, 19
timeline, 14-15
train, 14, 15, 16, 19
users, 6, 7, 8, 10, 16, 17
video games, 7, 16, 17
VIEW, 15

The images in this book are reproduced through the courtesy of: pickingpok, cover; Mark Nazh, CIP; Viewpoint Games, p. 4; Aleksandra Suzi, p. 5; DC Studio, p. 6; PR Image Factory, p. 7 (top left); Roman Zaiets, p. 7 (top right); NASA, p. 7 (bottom left); Niyazz, p. 7 (bottom right); Uladzimir Gudvin, p. 8; ferita Rahayuningsih, p. 9; Christian Bertrand, p. 10; Morton Heilig Collection/ Hugh M. Hefner Moving Image Archive/ USC School of Cinematic Arts, pp. 12, 14 (top left); Matthew Corley, p. 13; Dick Lyon, p. 14 (bottom); Sanjay Acharya, p. 14 (right); Matthew Paul Argall, p. 15 (bottom right); Kira_Yan, p. 15 (top); dennizn, p. 15 (bottom right); Daniel Fung, p. 16; https://teslasuit.io/, pp. 18, 19; Tinxi, p. 20; mpohodzhay, p. 21 (top left); Andrey_Popov, p. 21 (top right); ImageFlow, p. 21 (middle left); leungchopan, p. 21 (middle right); Y Photo Studio, p. 21 (bottom left); Rommel Canlas, p. 21 (bottom right).